Sidenotes For The Reason

By

Ian Beardsley

Copyright © 2014 by Ian Beardsley

ISBN: 978-1-312-32323-0

These are side notes to my book The Reason, which has the sequel The Reason: Moving On. In that book, since 9/5 was the ratio of Saturn orbit to Jupiter orbit in their closest approaches to the sun, while they are the largest and most massive planets in the solar system, mystically putting the earth at one unit from the sun, and further since 9 to 5 is the ratio of gold to silver in molar masses, and is the ratio of the solar radius to the lunar orbital radius, with the sun gold in color and the moon silver in color, and even further since 9 to 5 is 1.8 and I show 9 to 5 unifies pi with the golden ratio and pi with euler's number e, we formed three equations:

The Neptune Equation

The Uranus Equation

The Earth Equation

To find the normal vector given by the Sagittarius Equation, and we found it points to Sagittarius, near the origin of The SETI Wow! Signal, the one signal that the Search For Extraterrestrial Intelligence found that has the thumbprint of extraterrestrial origins. However, the Sagittarius normal comes from the equation of a plane, not from the equation of the plane. We will see what I mean by that in this work. We will derive the equation of the plane and we find the normal vector is in the constellation Scutum. The only difference between the normal vector of Sagittarius and the normal vector of Scutum, is that the y-component of the Scutum vector is the y-component of the Sagittarius vector increased by a factor of two.

The constellation Scutum is in the Mexican Skies between Aquila and Ophiucus. When I worked at Pine Mountain Observatory, in Central Oregon, Professor Kemp, who ran the observatory took me to the edge of the hill where we could see low on the horizon at sunrise, Ophiucus rising, probably because he grew up in Mexico and this was an opportunity to get a rare glimpse of a Southern Hemisphere constellation he was familiar with as a child. Aquila is the Eagle and is an ancient constellation known as the Eagle before Christ. In Greek Mythology, it was the eagle attendant to Zeus, or Jupiter (Roman). Scutum is in a particularly rich area of the Milky Way, full of dense star clouds. It has the Scutum Star Cloud. In 2006, NASA discovered Scutum to be a powerful source of x-rays and gamma rays. They found it was due to one of the most massive star clusters in the galaxy, and was caused by an unusually high number of red supergiants which were going supernova, which is the tendency for this type of star to end it life. Finally, we find the unusual occurrence of 0.621 in doing the Scutum calculation that occurs in my prediction for hyperdrive and the McKenna timewave as the novelty rating for the year humans landed on the moon reduced by ten, I call it here The Levinson Number. Read my work: Fiction-Reality Entanglement: Levinson, Asimov, And Clarke.

Ian Beardsley, June 27, 2014

$$x - 1 = t$$
$$y - 2 = t$$
$$z - 3 = t$$

$$(x - 1) - (y - 2) = 0$$
$$\boxed{x - y + 1 = 0}$$

$$(y - 2) - (z - 3) = 0$$
$$\boxed{y - z + 1 = 0}$$

$$x - y + 1 = y - z + 1$$
$$x - y - y + z = 0$$
$$\boxed{x - 2y + z = 0}$$

Equation of
the
Plane

$$x - 1 = y - 2 = z - 3$$
$$x - y + 1 = z - 3$$
$$\boxed{x - y - z + 2 = 0}$$

Equation of
a
Plane

①

$$\frac{5x}{36} + \frac{20}{36} = t$$

$$\frac{10y}{33} - \frac{30}{33} = t \qquad \text{Let} \quad t = 0$$

$$\frac{z}{9} + \frac{5}{9} = t$$

$$\frac{5}{36}x = -\frac{20}{36} \qquad x = -4$$

$$\frac{10}{33}y = \frac{30}{33} \qquad y = 3$$

$$\frac{1}{9}z = -\frac{5}{9} \qquad z = -\frac{5}{9}$$

$$7.2x = \frac{36}{5}x \qquad \text{inverted} \qquad \frac{5}{36}x$$

$$3.3x = \frac{33}{10}x \qquad \qquad \frac{10}{33}y$$

$$9x = 9x \qquad \qquad \frac{1}{9}z$$

②

7.2x

inversion
but not negative

$\frac{1}{7.2}x = \frac{5}{36}x$

$3.3x = \frac{33}{10}x$

negative
inversion
is perpendicular

$-\frac{10}{33}x = \frac{1}{-3.3}x$

9x

$-\frac{1}{9}x$

negative
inversion
is perpendicular

neoune equation   uranos equation   ③
(7.2) (3.3) earth equation
        (9)

$$\frac{5x}{36} \qquad -\frac{10}{33}y \qquad -\frac{1}{9}z \quad \text{coefficients}$$

$$\frac{1}{7.2} \qquad -\frac{1}{3.3} \qquad -\frac{1}{9} \quad \text{inverted coefficients}$$

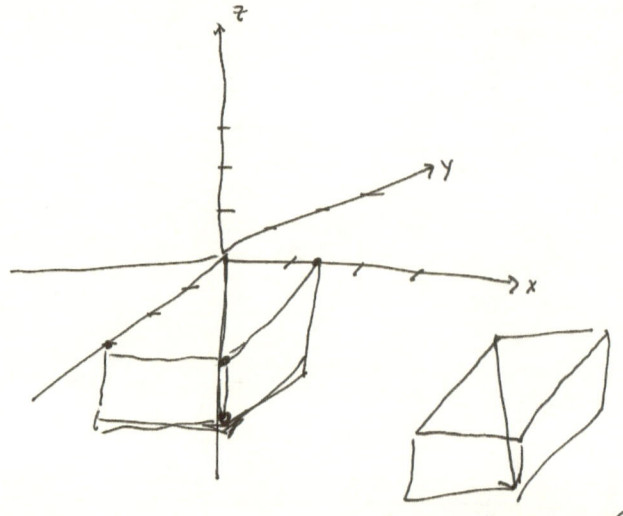

Sagittarios

$$\frac{5}{36} = 0.1389 \qquad -\frac{1}{9} = 0.11$$

$$-\frac{1}{3.3} = 0.30$$

For the Sagittarius equation
we considered the equation
of a plane. Let us now consider
the equation of the plane.

$$t = \frac{5x}{36} + 20 \qquad t = \frac{10y}{33} - 30 \qquad t = \frac{z}{9} + 5$$

$$\frac{5x}{36} + 20 = \frac{10y}{33} - 30$$

$$\boxed{\frac{5x}{36} - \frac{10y}{33} + 50 = 0}$$

$$\frac{5x}{36} + 20 = \frac{z}{9} + 5 \qquad \frac{10y}{33} - 30 = \frac{z}{9} + 5$$

$$\boxed{\frac{5x}{36} - \frac{z}{9} + 15 = 0} \qquad \frac{10y}{33} - 25 = \frac{z}{9}$$

$$\boxed{\frac{10y}{33} - \frac{z}{9} - 25 = 0}$$

$$\frac{5x}{36} - \frac{10y}{33} + 50 = \frac{10y}{33} - \frac{z}{9} - 25$$

$$\frac{5x}{36} - \frac{10y}{33} - \frac{10y}{33} - \frac{z}{9} + 75 = 0$$

$$\boxed{\frac{5}{36}x - \frac{20y}{33} - \frac{z}{9} + 75 = 0} \leftarrow$$

②

$$\frac{5}{36}x - \frac{20y}{33} - \frac{7}{9} + 75 = 0$$

$$\nabla f = \left\langle \frac{5}{36}, -\frac{20}{33}, -\frac{1}{9} \right\rangle$$

$$a = \frac{5}{36} \qquad b = -\frac{20}{33} \qquad d = -\frac{1}{9}$$

$$c = \sqrt{\left(\frac{5}{36}\right)^2 + \left(\frac{20}{33}\right)^2} =$$

$$\sqrt{0.01929 + 0.3673} = \sqrt{0.38659}$$

$$= 0.62176 \quad \leftarrow \text{Levinson number reduced} \\ \text{by factor of ten}$$

$$\tan \alpha = \frac{b}{a} \qquad \qquad (\text{see fiction – reality} \\ \text{entanglement})$$

$$= -\frac{20}{33} \div \frac{5}{36} = -\frac{48}{11} = -4.363636$$

$$\alpha = \arctan(4.363636) = -77°$$

$$\tan \beta = \frac{d}{c} = -\frac{1}{9} \div 0.62176 = -0.1787$$

$$\beta = \arctan(-0.1787) = -10.13 \approx -10°$$

③

α = right ascension

β = declination

$\alpha = -77°$

$$\frac{-77°}{15°/hour} = -5 \text{ hours}$$

24 hours $-$ 5 hours $=$ 19 hours

our star is located at:

RA: 19 hours

Dec: $-10°$

This is around the star

HD 176853 19^h 03m 25.05s

$-10°$ 41' 48.8"

In the constellation scutum

It has visual magnitud 6.64

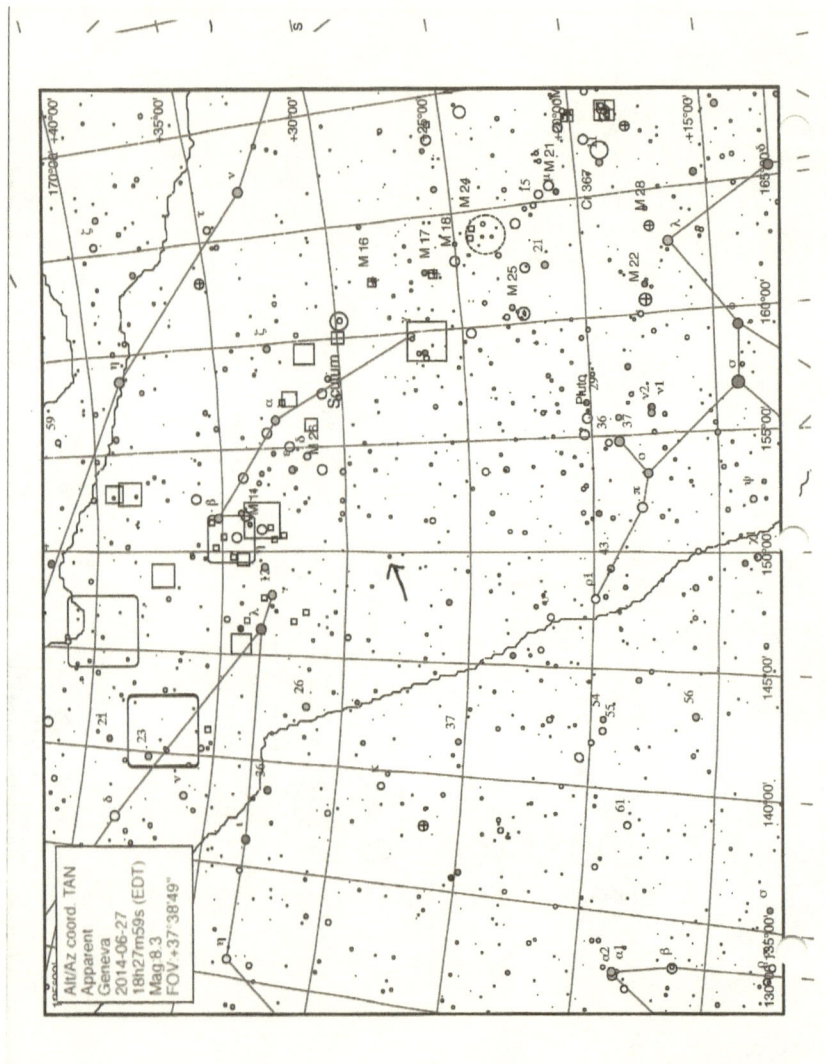

In my paper Paul Levinson, Isaac Asimov, Arthur C. Clarke Intertwined With An Astronomer's Research, I make the mathematical prediction that "humans have a 70% chance of developing Hyperdrive in the year 2043" to word it as Paul Levinson worded it, and I point out that this is only a year after the character Sierra Waters is handed a newly discovered document that sets in motion the novel by Paul Levinson, "The Plot To Save Socrates".

I now find that Isaac Asimov puts such a development in his science fiction at a similar time in the future, precisely in 2044, only a year after my prediction and two years after Sierra Waters is handed the newly discovered document that initiates her adventure. So, we have my prediction, which is related to the structure of the universe in a mystical way right in between the dates of Levinson and Asimov, their dates only being a year less and a year greater than mine.

Asimov places hyperdrive in the year 2044 in his short story "Evidence" which is part of his science fiction collection of short stories called, "I, Robot".

This is a collection of short stories where Robot Psychologist Dr. Susan Calvin is interviewed by a writer about her experience with the company on earth in the future that first developed sophisticated robots. In this book, the laws of robotics are created and the idea of the positronic brain introduced, and the nature of the impact robots would have on human civilization is explored. Following this collection of stories Asimov wrote three more novels, which comprise the robot series, "The Caves of Steel", "The Naked Sun", and "The Robots of Dawn".

"I, Robot" is Earth in the future just before Humanity settles the more nearby stars. The novels comprising "The Robot Series" are when humanity has colonized the nearby star systems, The Foundation Trilogy, and its prequels and sequels are about the time humanity has spread throughout the entire galaxy and made an Empire of it. All of these books can be taken together as one story, with characters and events in some, occurring in others.

Hyperdrive is invented in I, Robot by a robot called The Brain, owned by the company for which Dr. Susan Calvin works when it is fed the mathematical logistical problems of making hyperdrive, and asked to solve them. It does solve them and it offers the specs on building an interstellar ship, for which two engineers follow in its construction. They are humorously sent across the galaxy by The Brain, not expecting it, and brought back to earth in the ship after they constructed it. This was in the story in "I, Robot" titled "Escape!".

But Dr. Susan Calvin states in the following short story, that I mentioned, "Evidence":

"But that wasn't it, either"…"Oh, eventually, the ship and others like it became government property; the Jump through hyperspace was perfected, and now we actually have human colonies on the planets of some of the nearer stars, but that wasn't it."

"It was what happened to the people here on Earth in last fifty years that really counts."

And, what happened to people on Earth? The answer is in the same story "Evidence" from which that quote is at the beginning. It was when the Regions of the Earth formed The Federation. Dr. Susan Calvin says at the end of the story "Evidence":

"He was a very good mayor; five years later he did become Regional Co-ordinator. And when the Regions of Earth formed their Federation in 2044, he became the first World Co-ordinator."

It is from that statement that I get my date of 2044 as the year Asimov projects for hyperdrive.

Ian Beardsley
March 17, 2011

Wikipedia Encyclopedia writes:

""Timewave zero" is a numerological formula that purports to calculate the ebb and flow of "novelty", defined as increase in the universe's interconnectedness, or organised complexity,[66] over time. According to Terence McKenna, who conceived the idea over several years in the early- to mid-1970s while using psilocybin mushrooms and DMT, the universe has a teleological attractor at the end of time that increases interconnectedness, eventually reaching a singularity of infinite complexity in 2012, at which point anything and everything imaginable will occur simultaneously."

One could say Paul Levinson and Isaac Asimov could lay down the same card for the star Alpha Centauri, because that star was involved in my calculation for hyperdrive that intersected with The Plot To Save Socrates and I, Robot. A crossed destiny of a sort.

I watched a video on youtube about Terence McKenna where he lectured on his timewave zero theory. I found there was not an equation for his timewave zero graph but that a computer algorithm generated the graph of the wave. The next day I did a search on the internet to see if a person could download timewave software for free. As it turned out one could, for both Mac and pc. It is called "Timewave Calculator Version 1.0". I downloaded the software and found you had to download it every time after you quit the application and that you could not save the graph of your results or print them out. So I did a one-time calculation. It works like this: you input the range of time over which you want see the timewave and you cannot calculate past 2012, because that is when the timewave ends. You also put in a target date, the time when you want to get a rating for the novelty of the event that occurred on that day. You can also click on any point in the graph to get the novelty rating for that time. I put in:

Input:

Begin Date: December 27 1968 18 hours 5 minutes 37 seconds
End Date: December 2 2011 0 hours 28 minutes 7 seconds

McKenna said in the video on youtube that the dips, or valleys, in the timewave graph represent novelties. So, I clicked on the first valley after 1969 since that is the year we went to the moon, and the program gave its novelty as:

Sheliak Timewave Value For Target:

0.0621

On Target Date: August 4, 1969 9 hours 53 minutes 38 seconds

I was happy to see this because, I determined that the growth rate constant, k, that rate at which we progress towards hyperdrive, in my calculation in my work Asimovian Prediction For Hyperdrive, that gave the date 2043, a year after Sierra Waters was handed the newly discovered document that started her adventure in The Plot To Save Socrates, by Paul Levinson, and a year before Isaac Asimov had placed the invention of hyperdrive in his book I, Robot, was:

$(k=0.0621)$

The very same number!!!

In the case of Isaac Asimov, we are far in the future of humanity. In his Robot Series, Asimov has man making robots whose programming only allows them to do that which is good for humanity. As a result, these robots, artificial intelligence (AI), take actions that propel humanity into settling the Galaxy, in the robot series, and ultimately save humanity after they have settled the Galaxy and made an empire of it (In the Foundation Series).

In the case of Paul Levinson, scholars in the future travel through time and use cloning, a concept related to artificial intelligence (it is the creating of human replicas as well, but biological, not electronic), and the goal is to save great ancient thinkers from Greece, and to manipulate events in the past for a positive outcome for the future of humanity, just as the robots try to do in the work of Asimov.

In the case of Arthur C. Clarke, man undergoes a transformation due to a monolith placed on the moon and earth by extraterrestrials who have created life on earth. The monolith is a computer. It takes humans on a voyage to other planets in the solar system, and in their trials, humanity goes through trials that result in a transformation for the ending of their dependence on their technology and for becoming adapted to life in the Universe beyond Earth. That is, the character Dave Bowman becomes the Starchild in his mission to Jupiter. The artificial intelligence is the ship computer called HAL.

So, the thread is the salvation of man through technology, and their transformation to a new human paradigm, where they can end their dependence on Earth and adapt to the nature of the Universe as a whole.

At the time I was reading these novels, I was doing astronomical research, and, to my utter astonishment, my relationships I was discovering pertaining to the Universe were turning up times and values pivotal to these works of Levinson, Asimov, and Clarke. Further, I was interpreting much of my discoveries by developing them in the context of short fictional stories.

It is a curious thing that the Earth is the third planet from the Sun and the third brightest star in the sky is the closest to us and very similar to the sun in a galaxy of a rich variety of stars. This closest star to us is a triple system known as Alpha Centauri A, B, and Proxima Centauri. Alpha Centauri A is, like our Sun a main sequence spectral type G star. Precisely, G2 V, just as is the Sun. Its physical characteristics are very close to those of the Sun: 1.10 solar masses, 1.07 solar diameters, and 1.5 solar luminosities. It is absolute magnitude +4.3. The absolute magnitude of the Sun is +4.83.

If ever the option existed for humans to travel to the stars, this situation speaks of it, whether or not Alpha Centauri has an earth-like planet in its habitable zone.

It has been said that the base ten place significant system of writing numbers stems from the fact we have ten fingers to count on. In so far as science can save us, it can destroy us in that science is not dangerous, but humans can be.

Traveling to the planets is possible with chemical fuel rockets, but traveling to the stars is another story, because of their immense distances from us, and from one another.

What are the odds that our development in technologies will take us to the stars before we destroy ourselves first? In other words, what are the intrinsic odds for humankind to develop the hyperdrive before without bringing about its own end first?

We do a random walk to Alpha Centauri of 10 one light year jumps. We make 10 equal steps randomly of one light year each, equal steps that if all are towards Alpha Centauri we will land beyond it. If 10 are away from it, we are as far from it as can be. And, if 5 are towards it, and five are away from it, we have gone nowhere.

In this allegory we calculate the probability of landing on Alpha Centauri, in 10 random leaps of a light year each, a light year being the distance light travels in the time it takes the earth to make one revolution around the sun, light speed a natural constant.

I have pointed out that $0.0621 = k$ the growth rate constant towards hyperdrive in my paper, The Levinson,-Asmov-Clarke Phenomenon. I have pointed out that in McKenna time wave theory 0.0621 is is the novelty rating for 1969, the year man landed on the moon. Our counting system is base 10, it has been said probably because we have ten fingers. If we multiply 0.0621 by 10, that is increase it by a factor of ten, we get 0.621. If we round that to two places after the decimal, it is 0.62. Let us consider the golden ratio conjugate. The golden ratio occurs throughout nature. It is in the rotation from leaf to leaf around the stem of a plant, for example. The golden ratio conjugate is just the inverse of the golden ratio. It is simply the separation between leaves around the stem of a plant, going in the other direction. It is equal to 0.618 to three places after the decimal. Let us round that to two places after the decimal. The eight rounds the one to two. The golden ratio conjugate is then 0.62 rounded to two places after the decimal. That is the same value as k increased by a factor of 10, and the same value as the novelty rating for 1969 increased by a factor of ten. I find that interesting.

(continue to the next page)

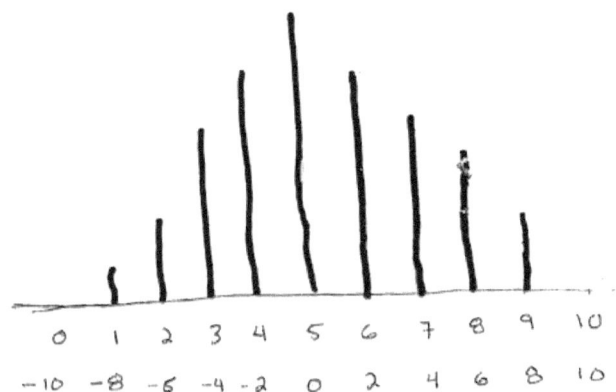

The probability of making n steps in either direction forms a bell shaped curve. After 10 randomly made steps the odds of going nowhere is highest and, is represented by five in the bell curve corresponding to 0. Let us round the distance of Alpha Centauri to four light years, giving humans the benefit of the doubt. The number positive four in the bell graph has written above it the number 7. Seven out of ten times 100 for effort gives a 70% chance of making it to the stars without becoming extinct first. I believe the percent understanding of our technological development towards hyperdrive, where we have just entered space with chemical rockets and developed fast, compact, computers, is given by:

$$W_N(n_1) = \frac{N!}{n_1! n_2!} p^{n_1} q^{n_2}$$

Evaluated at n1=7.

N is 10 steps.

And n1 is the number of steps towards Alpha Centauri, n2 those away from it.

And, p is the probability that the step is towards Alpha Centauri, and q is the probability that the step is away from Alpha Centauri.

N = n1 + n2

And m = n1-n2 is the displacement

And q+p=1

The trick to using this equation is in knowing the possible combinations of steps that can be made that equal 10. Like five right, five left with a displacement of 0 or, 10 right, 0 left with a displacement of 10 or, 7 left, 3 right with a displacement of negative 4.

To land at 4 light years from earth, with 10 one light year jumps, one must go away from Alpha Centauri 3 jumps of a light year each then 7 jumps toward it of one light year each, to land on it, that is to land at +4, its location. So n1 is 7 and n2 is 3. The probability to jump away from the star is 1/2 and the probability to jump towards it is 1/2. That is p=1/2 and q=1/2. There are ten random jumps, so, N=10.

Using our equation:

$$\frac{(10!)}{(7!)(3!)}(\frac{1}{2})^7(\frac{1}{2})^3 = \frac{3628800}{(5040)(6)}\frac{1}{128}\frac{1}{8} = \frac{120}{1024} = \frac{15}{128} = 0.1171875 \approx 12\%$$

We would be, by this reasoning 12% along in the development towards hyperdrive.

Ian Beardsley
June 2009

If human technology has ever been anything, it has been exponential, growing in proportion to itself. In other words, two developments beget 8, eight beget 16, and sixteen begets 32. My grandfather rode a horse when he was a child, as a young man he drove a car, and when I knew him as a child, he saw humans land on the moon.

It wasn't long before we made computers small enough that people could keep in their homes that did more than computers did in the 60's that filled an entire room.

Having calculated that we are 12% along in developing the hyperdrive, we can use the equation for natural growth to estimate when we will have hyperdrive. It is of the form:

$$x(t) = x_0 e^{kt}$$

t is time and k is a growth rate constant which we must determine to solve the equation. In 1969 Neil Armstrong became the first man to walk on the moon. In 2009 the European Space Agency launched the Herschel and Planck telescopes that will see back to near the beginning of the universe. 2009-1969 is 40 years. This allows us to write:

$$12\% = e^{k(40)}$$

$$\log 12 = 40k \log 2.718$$

$$0.026979531 = 0.4342 \, k$$

$$k = 0.0621$$

We now can write:

$$x(t) = e^{(0.0621)t}$$

$$100\% = e^{(0.0621)t}$$

$$\log 100 = (0.0621) \, t \log e$$

$$t = 74 \text{ years}$$

$$1969 + 74 \text{ years} = 2043$$

Our reasoning would indicate that we will have hyperdrive in the year 2043.

Study summary:

1. We have a 70% chance of developing hyperdrive without destroying ourselves first.
2. We are 12% along the way in development of hyperdrive.
3. We will have hyperdrive in the year 2043, plus or minus.

Sierra Waters was handed the newly discovered document in 2042.

Chapter 1

AE-35

I wrote a short story last night, called Gypsy Shamanism and the Universe about the AE-35 unit, which is the unit in the movie and book 2001: A Space Odyssey that HAL reports will fail and discontinue communication to Earth. I decided to read the passage dealing with the event in 2001 and HAL, the ship computer, reports it will fail in within 72 hours. Strange, because Venus is the source of 7.2 in my Neptune equation and represents failure, where Mars represents success.

Ian Beardsley
August 5, 2012

It must have been 1989 or 1990 when I took a leave of absence from The University Of Oregon, studying Spanish, Physics, and working at the state observatory in Oregon -- Pine Mountain Observatory—to pursue flamenco in Spain.

The Moors, who carved caves into the hills for residence when they were building the Alhambra Castle on the hill facing them, abandoned them before the Gypsies, or Roma, had arrived there in Granada Spain. The Gypsies were resourceful enough to stucco and tile the abandoned caves, and take them up for homes.

Living in one such cave owned by a gypsy shaman, was really not a down and out situation, as these homes had plumbing and gas cooking units that ran off bottles of propane. It was really comparable to living in a Native American adobe home in New Mexico.

Of course living in such a place came with responsibilities, and that included watering its gardens. The Shaman told me: "Water the flowers, and, when you are done, roll up the hose and put it in the cave, or it will get stolen". I had studied Castilian Spanish in college and as such a hose is "una manguera", but the Shaman called it "una goma" and goma translates as rubber. Roll up the hose and put it away when you are done with it: good advice!

So, I water the flowers, rollup the hose and put it away. The Shaman comes to the cave the next day and tells me I didn't roll up the hose and put it away, so it got stolen, and that I had to buy him a new one.

He comes by the cave a few days later, wakes me up asks me to accompany him out of The Sacromonte, to some place between there and the old Arabic city, Albaicin, to buy him a new hose.

It wasn't a far walk at all, the equivalent of a few city blocks from the caves. We get to the store, which was a counter facing the street, not one that you could enter. He says to the man behind the counter, give me 5 meters of hose. The man behind the counter pulled off five meters of hose from the spindle, and cut the hose to that length. He stated a value in pesetas, maybe 800, or so, (about eight dollars at the time) and the Shaman told me to give that amount to the man behind the counter, who was Spanish. I paid the man, and we left.

I carried the hose, and the Shaman walked along side me until we arrived at his cave where I was staying. We entered the cave stopped at the walk way between living room and kitchen, and he said: "follow me". We went through a tunnel that had about three chambers in the cave, and entered one on our right as we were heading in, and we stopped and before me was a collection of what I estimated to be fifteen rubber hoses sitting on ground. The Shaman told me to set the one I had just bought him on the floor with the others. I did, and we left the chamber, and he left the cave, and I retreated to a couch in the cave living room.

Chapter 2

Gypsies have a way of knowing things about a person, whether or not one discloses it to them in words, and The Shaman was aware that I not only worked in Astronomy, but that my work in astronomy involved knowing and doing electronics.

So, maybe a week or two after I had bought him a hose, he came to his cave where I was staying, and asked me if I would be able to install an antenna for television at an apartment where his nephew lived.

So this time I was not carrying a hose through The Sacromonte, but an antenna.

There were several of us on the patio, on a hill adjacent to the apartment of The Shaman's Nephew, installing an antenna for television reception.

Chapter 3

I am now in Southern California, at the house of my mother, it is late at night, she is a asleep, and I am about 24 years old and I decide to look out the window, east, across The Atlantic, to Spain. Immediately I see the Shaman, in his living room, where I had eaten a bowl of the Gypsy soup called Puchero, and I hear the word Antenna. I now realize when I installed the antenna, I had become one, and was receiving messages from the Shaman.

The Shaman's Children were flamenco guitarists, and I learned from them, to play the guitar. I am now playing flamenco, with instructions from the shaman to put the gypsy space program into my music. I realize I am not just any antenna, but the AE35 that malfunctioned aboard The Discovery just before it arrived at the planet Jupiter in Arthur C. Clarke's and Stanley Kubrick's "2001: A Space Odyssey". The Shaman tells me, telepathically, that this time the mission won't fail.

Chapter 4

I am watching Star Wars and see a spaceship, which is two oblong capsules flying connected in tandem. The Gypsy Shaman says to me telepathically: "Dios es una idea: son dos". I understand that to mean "God is an idea: there are two elements". So I go through life basing my life on the number two.

Chapter 5

Once one has tasted Spain, that person longs to return. I land in Madrid, Northern Spain, The Capitol. The Spaniards know my destination is Granada, Southern Spain, The Gypsy Neighborhood called The Sacromonte, the caves, and immediately recognize I am under the spell of a Gypsy Shaman, and what is more that I am The AE35 Antenna for The Gypsy Space Program. Flamenco being flamenco, the Spaniards do not undo the spell, but reprogram the instructions for me, the AE35 Antenna, so that when I arrive back in the United States, my flamenco will now state their idea of a space program. It was of course, flamenco being flamenco, an attempt to out-do the Gypsy space program.

Chapter 6

I am back in the United States and I am at the house of my mother, it is night time again, she is asleep, and I look out the window east, across the Atlantic, to Spain, and this time I do not see the living room of the gypsy shaman, but the streets of Madrid at night, and all the people, and the word Jupiter comes to mind and I am about to say of course, Jupiter, and The Spanish interrupt and say "Yes, you are right it is the largest planet in the solar system, you are right to consider it, all else will flow from it."

I know ratios, in mathematics are the most interesting subject, like pi, the ratio of the circumference of a circle to its diameter, and the golden ratio, so I consider the ratio of the orbit of Saturn (the second largest planet in the solar system) to the orbit of Jupiter at their closest approaches to The Sun, and find it is nine-fifths (nine compared to five) which divided out is one point eight (1.8).

I then proceed to the next logical step: not ratios, but proportions. A ratio is this compared to that, but a proportion is this is to that as this is to that. So the question is: Saturn is to Jupiter as what is to what? Of course the answer is as Gold is to Silver. Gold is divine; silver is next down on the list. Of course one does not compare a dozen oranges to a half dozen apples, but a dozen of one to a dozen of the other, if one wants to extract any kind of meaning. But atoms of gold and silver are not measured in dozens, but in moles. So I compared a mole of gold to a mole of silver, and I said no way, it is nine-fifths, and Saturn is indeed to Jupiter as Gold is to Silver.

I said to myself: How far does this go? The Shaman's son once told me he was in love with the moon. So I compared the radius of the sun, the distance from its center to its surface to the lunar orbital radius, the distance from the center of the earth to the center of the moon. It was Nine compared to Five again!

Chapter 7

I had found 9/5 was at the crux of the Universe, but for every yin there had to be a yang. Nine fifths was one and eight-tenths of the way around a circle. The one took you back to the beginning which left you with 8 tenths. Now go to eight tenths in the other direction, it is 72 degrees of the 360 degrees in a circle. That is the separation between petals on a five-petaled flower, a most popular arrangement. Indeed life is known to have five-fold symmetry, the physical, like snowflakes, six-fold. Do the algorithm of five-fold symmetry in reverse for six-fold symmetry, and you get the yang to the yin of nine-fifths is five-thirds.

Nine-fifths was in the elements gold to silver, Saturn to Jupiter, Sun to moon. Where was five-thirds? Salt of course. "The Salt Of The Earth" is that which is good, just read Shakespeare's "King Lear". Sodium is the metal component to table salt, Potassium is, aside from being an important fertilizer, the substitute for Sodium, as a metal component to make salt substitute. The molar mass of potassium to sodium is five to three, the yang to the yin of nine-fifths, which is gold to silver. But multiply yin with yang, that is nine-fifths with five-thirds, and you get 3, and the earth is the third planet from the sun.

I thought the crux of the universe must be the difference between nine-fifths and five-thirds. I subtracted the two and got two-fifteenths! Two compared to fifteen! I had bought the Shaman his fifteenth rubber hose, and after he made me into the AE35 Antenna one of his first transmissions to me was: "God Is An Idea: There Are Two Elements".

It is so obvious, the most abundant gas in the Earth Atmosphere is Nitrogen, chemical symbol 15!

Chapter 8: The Sequence

We considered the ratio nine to five, then the proportion and found it in Saturn Orbit to Jupiter orbit, Solar Radius to Lunar Orbit, Gold to Silver and if flower petal arrangements. It is left then to consider the whole number multiples of nine-fifths (1.8) or the sequence:

1.8, 3.6, 5.4, 7.2,...

in other words, and we look to see if it is in the solar system and find it is in the following ways:

1.8

Saturn Orbit/Jupiter Orbit
Solar Radius/Lunar Orbit
Gold/Silver

3.6

(10)Mercury Radius/Earth Radius
(10)Mercury Orbit/Earth Orbit

(earth radius)/(moon radius)=
4(degrees in a circle)(moon distance)/(sun distance)
= 3.7 ~ 3.6

There are about as many days in a year as degrees in a circle.

(Volume of Saturn/Volume Of Jupiter)(Volume Of Mars) = 0.37 cubic earth radii
~ 3.6

The latter can be converted to 3.6 by multiplying it by (Earth Mass/Mars Mass) because Earth is about ten times as massive as Mars.

5.4

Jupiter Orbit/Earth Orbit
Saturn Mass/Neptune Mass

7.2

10(Venus Orbit/Earth Orbit)

Chapter 9: The Neptune Equation

If we consider as well the sequence where we begin with five and add nine to each successive term: 5, 14, 23, 32…Then, the structure of the solar system and dynamic elements of the Universe and Nature in general are tied up in the two sequences:

5, 14, 23, 32,…

and

1.8, 3.6, 5.4, 7.2,…

How do we find the connection between the two to localize the pivotal point of the solar system? We take their difference, subtracting respective terms in the second sequence from those in the first sequence to obtain the new sequence:

3.2, 10.4, 17.6, 24.8,…

Which is an arithmetic sequence with common difference of 7.2 meaning it is written

$7.2n - 4 = a_n$

The a_n is the nth term of the sequence, n is the number of the term in the sequence.

This we notice can be written:

[(Venus-orbit)/(Earth-orbit)][(Earth-mass)/(Mars-mass)]n – (Mars orbital #) = a_n

We have an equation for a sequence that shows the Earth straddled between Venus and Mars. Venus is a failed Earth. Mars promises to be New Earth.

The Mars orbital number is 4. If we want to know what planet in the solar system holds the key to the success of Earth, or to the success of humans, we let n =3 since the Earth is the third planet out from the Sun, in the equation and the result is a_n = 17.6. This means the planet that holds the key is Neptune. It has a mass of 17.23 earth masses, a number very close to our 17.6.

Not only is Neptune the indicated planet, we find it has nearly the same surface gravity as earth and nearly the same inclination to its orbit as earth. Though it is much more massive than earth, it is much larger and therefore less dense. That was why it comes out to have the same surface gravity.

Chapter 10: The Uranus Equation

I asked what needs to be done to solve My Neptune Equation, by going deep with the guitar in Solea Por Buleras. I found the answer was that I didn't have enough information to solve it.

Then I realized I could create the complement of the Neptune equation by looking at the Yang of 5/3, since the Neptune equation came from the Yin of 9/5.

We use the same method as for the Neptune equation:

Start with 8 and add 5 to each additional term (we throw a twist by not starting with 5)

5/3 => 8, 13, 18, 23,...

List the numbers that are whole number multiples of 5/3:

5/3n = 1.7, 3.3, 5, 6.7,...

Subtract respective terms in the second sequence from those in the first:

6.3, 9.7, 13, 16.3,...

This is an arithmetic sequence with common difference 3.3. It can be written:

$(a\_n) = 3 + 3.3n$

This can be wrtten:

Earth Orbital # + (Jupiter Mass/Saturn Mass)n = a_n

Letting n = 3 we find a_n = 13

The closest to this is the mass of Uranus, which is 14.54 earth masses. If Neptune is the Yin planet, then Uranus is the Yang planet. This is interesting because I had found that Uranus and Neptune were different manifestations of the same thing. I had written:

I calculate that though Neptune is more massive than Uranus, its volume is less such that their products are close to equivalent. In math:

N_v = volume of Neptune
N_m = mass of Neptune
U_v = volume of Uranus
U_m = mass of Uranus

$(N\_v)(N\_m) = (U\_v)(U\_m)$

Chapter 11: The Earth Equation

We then sought the Yang of six-fold symmetry because it is typical to physical nature, like snowflakes. We said it was 5/3 since it represents the 120 degree measure of angles in a regular hexagon and we built our universe from there, resulting in the Uranus Integral, which was quite fruitful. Let us, however, think of Yang not as 5/3, but look at the angles between radii of a regular hexagon. We have:

$360 - 60 = 300$

$300 + 360 = 660$

$660/360 = 11/6$

We say Yin is 9/5 and Yang is 11/6 and stick with The Gypsy Shaman's 15 (See An Extraterrestrial Analysis, chapter titled "Gypsy Shamanism And The Universe") and build our Cosmology from there.

We already built The Neptune Equation from 9/5 and used it with 5/3 to derive the planet Europia, but let us apply 11/6 in place of 5/3:

$11/6 \Rightarrow 11/6, 11/3, 11/2, 22/3, \ldots = 1.833, 3.667, 5.5, 7.333, \ldots$

$11/6 \Rightarrow 6, 6+11 = 17, 17+11=28, 28+11=39, \ldots = 6, 17, 28, 39, \ldots$
Subtract the second sequence from the first:
$4.167, 13.333, 22.5, 31.667, \ldots$
Now we find the common difference between terms in the latter: $9.166, 9.167, 9.167, \ldots$

$(a\_n) = a + (n-1)d = 4.167+(n-1)9.167 = 4.167 + 9.167n - 9.167 = 9.167n-5$

Try n=3: $9.167(3) - 5 = 27.501 - 5 = 22.501$ (works)
Our equation is:

$(a\_n) = 9.167n - 5$

We notice this can be written:

[(Saturn Orbit)/(Earth Orbit)]n − (Jupiter Orbital #) = (a_n)

The Neptune Equation for n=3 gave Neptune masses, the Uranus equation for n=3 gave Uranus masses. This equation for n=3 gives close to the tilt of the Earth (23.5 degrees) in a form that is exactly half of the 45 degrees in a square with its diagonal drawn in. In the spirit of our first cosmology built upon 9/5, 5/3, and 15, we will call this equation The Earth Equation.

Chapter: 12: The Unification Of Pi and Phi by Nine-Fifths

All that is left to do is to consider pi, the circumference of a circle to its diameter, and phi, the golden ratio, since they are two most important, if not most beautiful ratios in mathematics.

I have found nine-fifths occurs throughout nature in the rotation of petals around a a flower for a most popular arrangement, in the orbits of jupiter to saturn in their closest approaches to the sun, in the ratio of the molar masses of gold to silver, and in the ratio of the solar radius to the lunar orbit. I now further go on to say that this nine-fifths unifies the two most important ratios in mathematics pi and the golden ratio (phi), in that

pi + phi = 3.141 + 1.618 = 4.759

Because the numbers after the decimal in the sum (the important part) are 5 and 9 and 7, the average of 5 and nine.

I should also like to point out that the fourth and fifth numbers after the decimal in pi are 5 and 9 and in phi the second and third numbers after the decimal are one and eight where nine-fifths divided out is one point eight, and, further, the first and second numbers after the decimal in phi add up to make 7, the average of 9 and 5, and subtract to make five, and the second and third digits after the decimal add up to nine. So not only does the solar system unify pi and phi through nine-fifths, pi and phi taken alone express nine-fifths in the best possible ways.

The Author

www.ingramcontent.com/pod-product-compliance
Lightning Source LLC
Chambersburg PA
CBHW021854170526
45157CB00006B/2439